MATHEMATICS

What you didn't know on your kindergarten math journey.

Let's have fun!

Presented by

Dr. Jannell Pearson-Campbell, Ed.D

Getting R*eady to Explore*

Always remember to have fun!
Create Pictures
Use items as visuals
Create a math journal for future use!
Every day Learn something new!
You can do it!!!

Table of Contents

Topic	PAGE
Counting Numbers from 0-10	5
Flip Chart Activity	68
Counting Up to 100 with Naming Each Number	79
Count to 100	90
Count by 5	92
Count by 10	95
One More	98
One Less	100
Money	102
Addition	104
Subtraction	106
Example of a Calendar and Days of the Week	108

Page 5

Understanding the Number

In Words	Symbol	Count
Zero		

Now It's Your Turn

In Words	Symbol	Count
Zero		
Zero		

Now It's Your Turn

In Words	Symbol	Count
Zero		
Zero		

Now It's Your Turn

In Words	Symbol	Count
Zero		
Zero		

Now It's Your Turn

In Words	Symbol	Count
Zero		
Zero		

Now It's Your Turn

In Words	Symbol	Count
Zero		
Zero		

Understanding the Number

In Words	Symbol	Count
One		

Now It's Your Turn

In Words	Symbol	Count
One		
one	1	

Now It's Your Turn

In Words	Symbol	Count
One		
one		

Now It's Your Turn

In Words	Symbol	Count
One		
one	1	

Now It's Your Turn

In Words	Symbol	Count
One		
one		

Understanding the number

In Words	Symbol	Count
Two		

Page 17

Not it's Your Turn!

In Words	Symbol	Count
Two	2	
Two	2	

Not it's Your Turn!

In Words	Symbol	Count
Two	2	
Two	2	

Not it's Your Turn!

In Words	Symbol	Count
Two	2	
Two	2	

Not it's Your Turn!

In Words	Symbol	Count
Two	2	
Two	2	

Not it's Your Turn!

In Words	Symbol	Count
Two	2	
Two	2	

Not it's Your Turn!

In Words	Symbol	Count
Two	2	
Two	2	

Understanding the number

In Words	Symbol	Count
Three		

Not it's Your Turn!

In Words	Symbol	Count
Three	3	
Three	3	

Not it's Your Turn!

In Words	Symbol	Count
Three	3	
Three	3	

Not it's Your Turn!

In Words	Symbol	Count
Three	3	
Three	3	

Not it's Your Turn!

In Words	Symbol	Count
Three	3	
Three	3	

Understanding the number

In Words	Symbol	Count
Four		

Not it's Your Turn!

In Words	Symbol	Count
Four		
Four		

Page 30

Not it's Your Turn!

In Words	Symbol	Count
Four		
Four		

Page 31

Not it's Your Turn!

In Words	Symbol	Count
Four		
Four		

Not it's Your Turn!

In Words	Symbol	Count
Four		
Four		

Page 33

Understanding the number

In Words	Symbol	Count
Five		

Not it's Your Turn!

In Words	Symbol	Count
Five	5	
Five	5	

Page 35

Not it's Your Turn!

In Words	Symbol	Count
Five	5	
Five	5	

Page 36

Not it's Your Turn!

In Words	Symbol	Count
Five	5	
Five	5	

Not it's Your Turn!

In Words	Symbol	Count
Five	5	
Five	5	

Page 38

Understanding the number

In Words	Symbol	Count
Six		

Not it's Your Turn!

In Words	Symbol	Count
Six	6	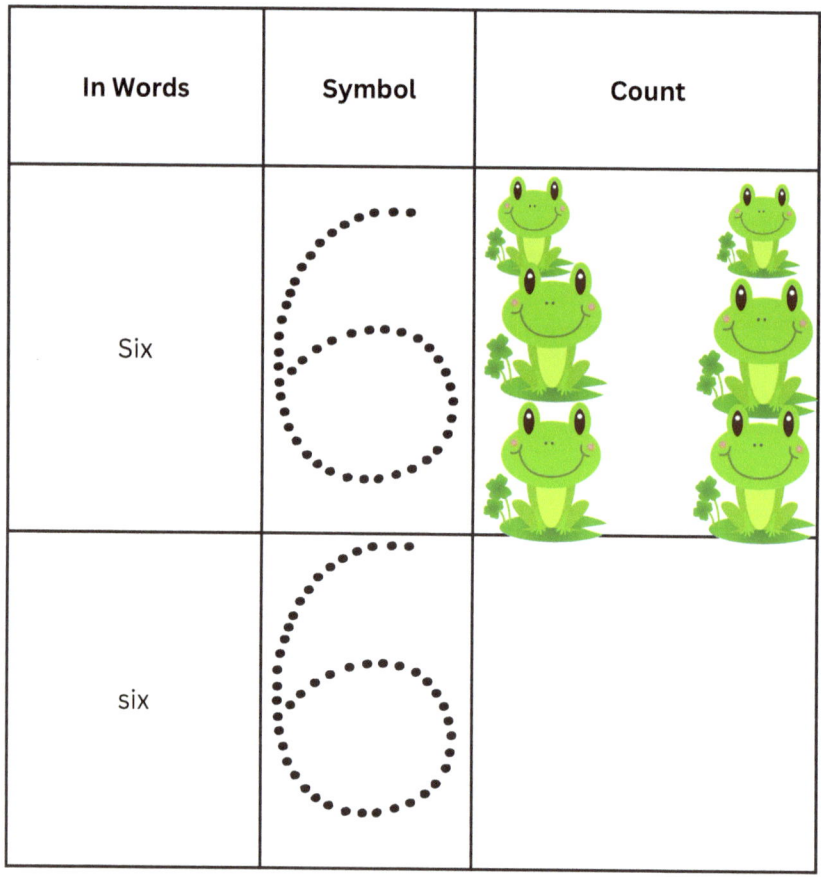
six	6	

Not it's Your Turn!

In Words	Symbol	Count
Six	6	
six	6	

Not it's Your Turn!

In Words	Symbol	Count
Six	6	
six	6	

Not it's Your Turn!

In Words	Symbol	Count
Six	6	
six	6	

Understanding the number

In Words	Symbol	Count
Seven	7	

Not it's Your Turn!

In Words	Symbol	Count
Seven	7	
Seven	7	

Not it's Your Turn!

In Words	Symbol	Count
Seven	7	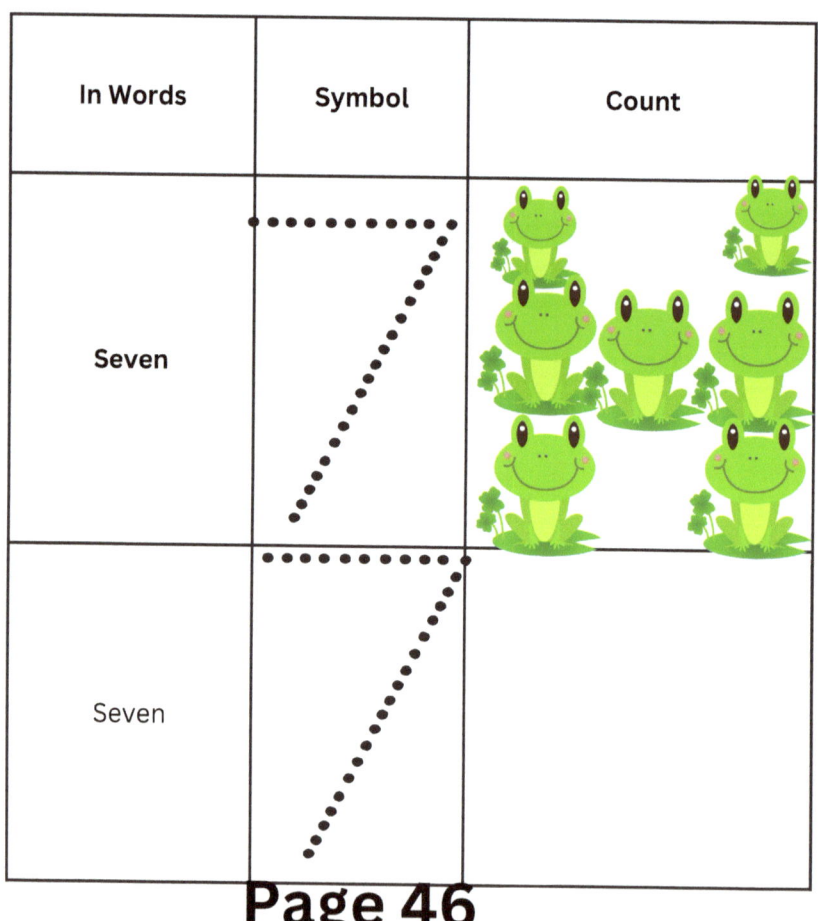
Seven	7	

Not it's Your Turn!

In Words	Symbol	Count
Seven	7	
Seven	7	

Not it's Your Turn!

In Words	Symbol	Count
Seven	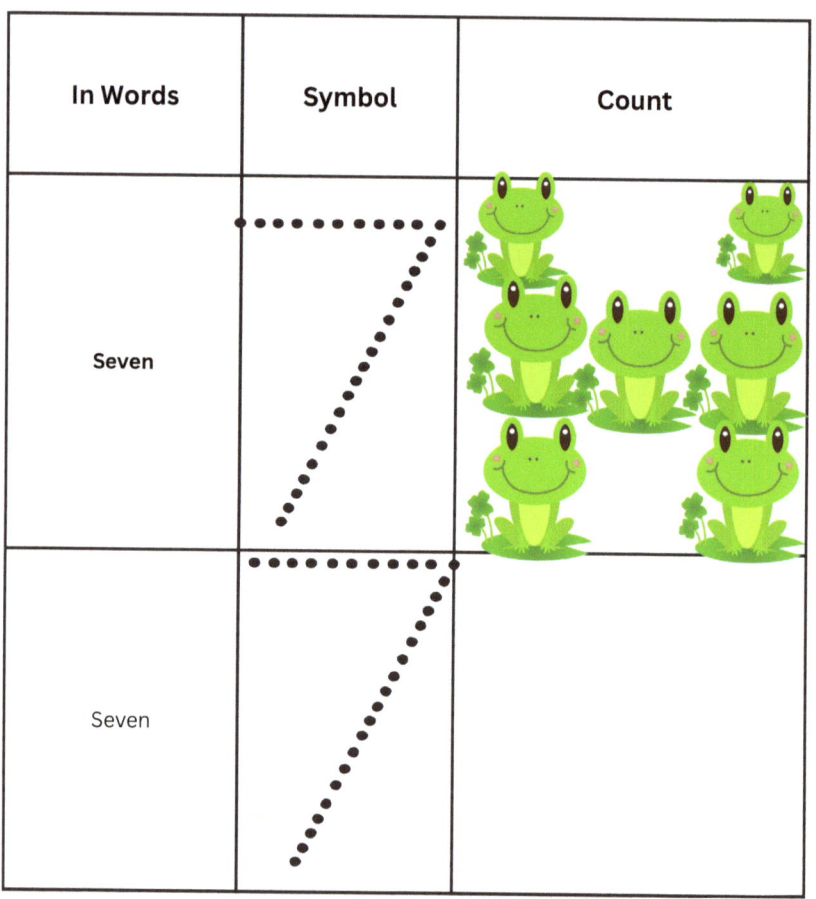	
Seven		

Understanding the number

In Words	Symbol	Count
Eight		

Not it's Your Turn!

In Words	Symbol	Count
Eight		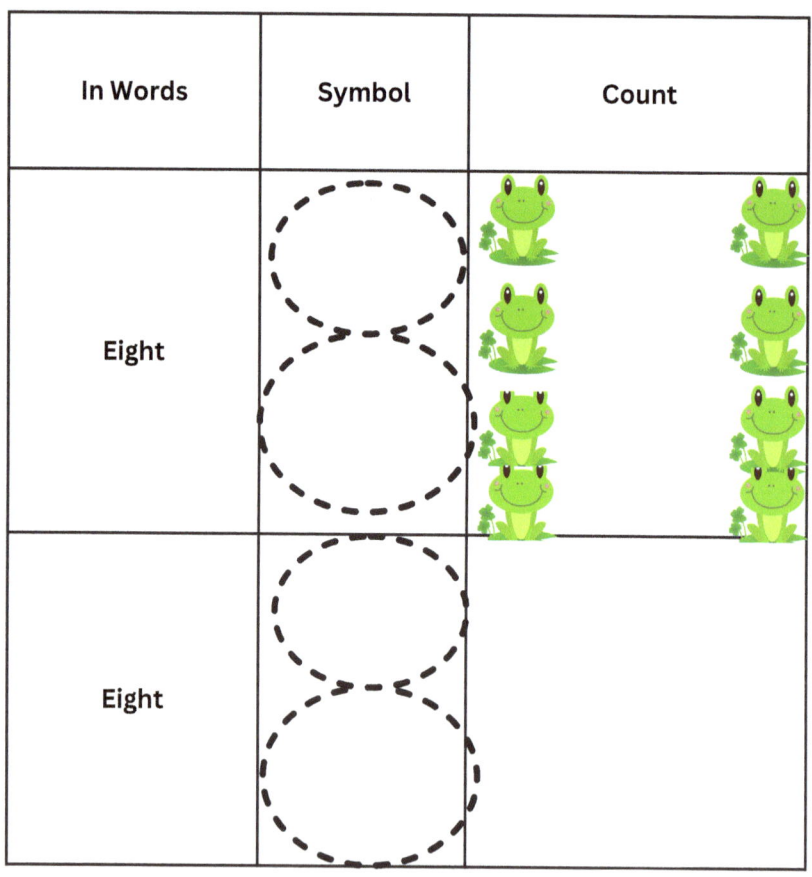
Eight		

Page 50

Not it's Your Turn!

In Words	Symbol	Count
Eight		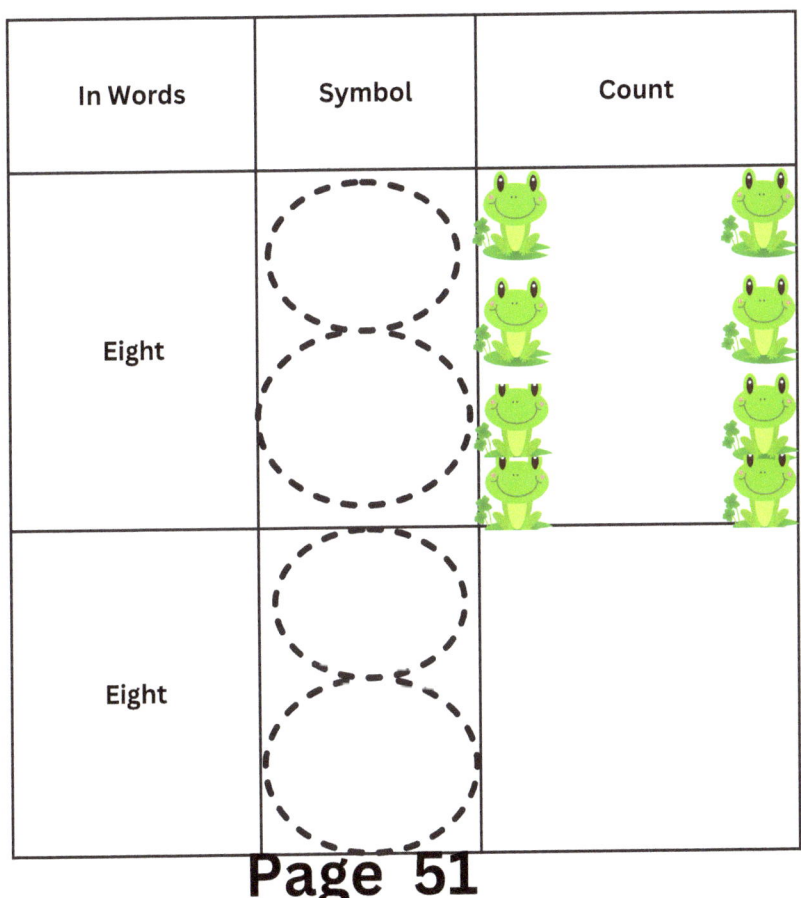
Eight		

Page 51

Not it's Your Turn!

In Words	Symbol	Count
Eight	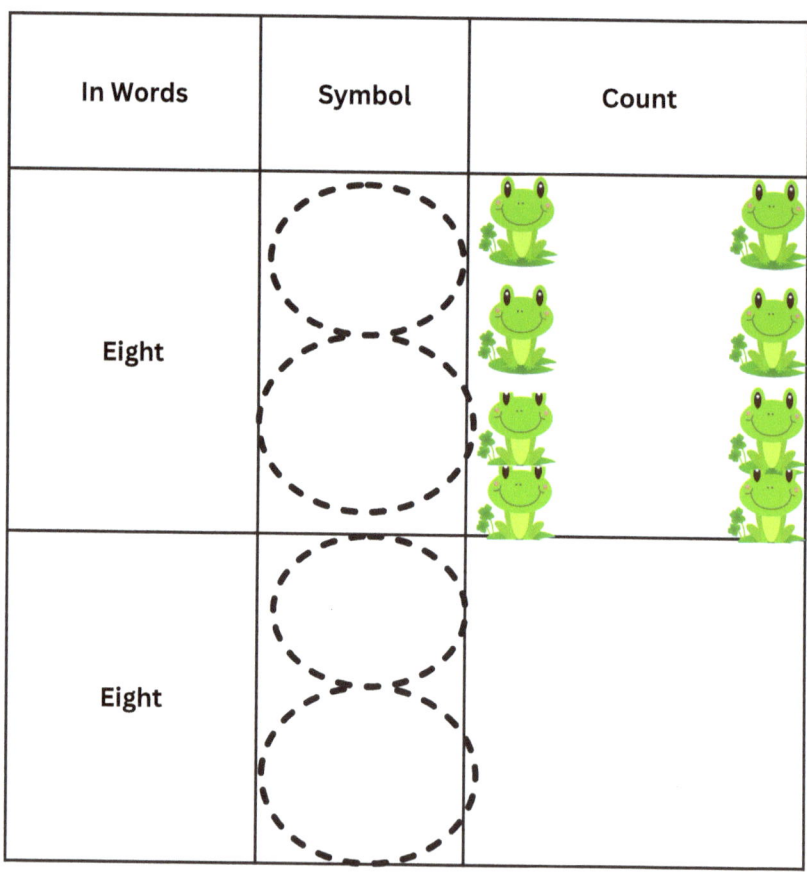	
Eight		

Not it's Your Turn!

In Words	Symbol	Count
Eight	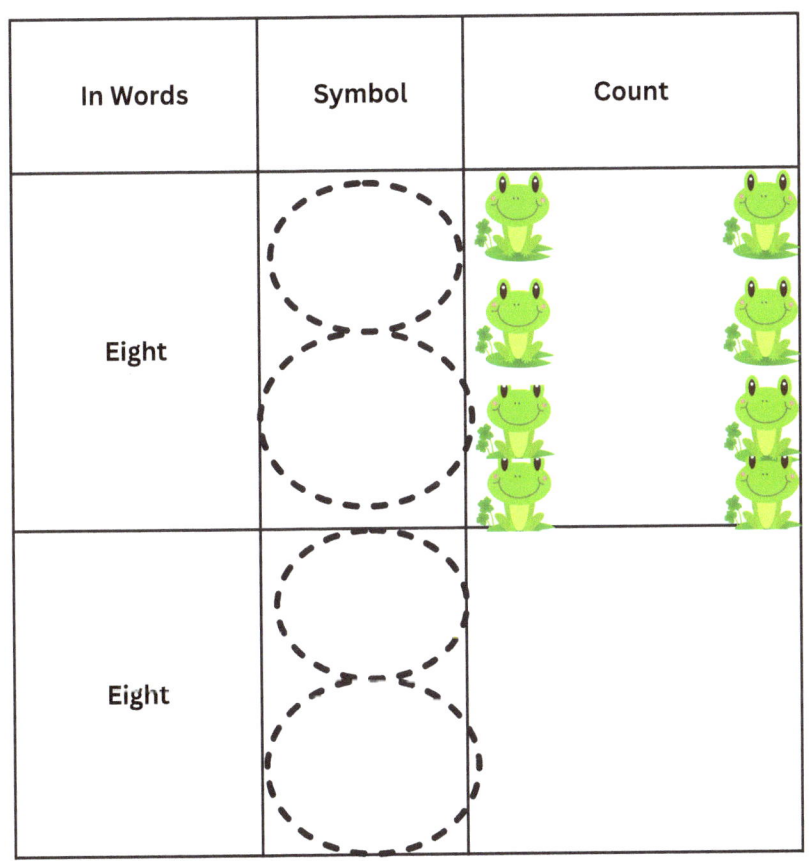	
Eight		

Understanding the number

In Words	Symbol	Count
Nine		

Not it's Your Turn!

In Words	Symbol	Count
Nine		
Nine		

Not it's Your Turn!

In Words	Symbol	Count
Nine		
Nine		

Not it's Your Turn!

In Words	Symbol	Count
Nine		
Nine		

Not it's Your Turn!

In Words	Symbol	Count
Nine		
Nine		

Page 58

Understanding the number

In Words	Symbol	Count
Ten	10	

Not it's Your Turn!

In Words	Symbol	Count
Ten	10	
Ten	10	

Not it's Your Turn!

In Words	Symbol	Count
Ten	10	
Ten	10	

Not it's Your Turn!

In Words	Symbol	Count
Ten	10	
Ten	10	

Not it's Your Turn!

In Words	Symbol	Count
Ten	10	
Ten	10	

Count to 10	
0	
1	
2	
3	
4	
5	
6	
7	
8	
9	
10	

Page 64

Page 65

Page 66

Page 67

Flip Chart Activity

Step 1 : Create 2 sets of cards numbering them 0-9

Step 2: Place the two cards next to each other

0

9

Step 3:

Place both cards together and the answer is 09

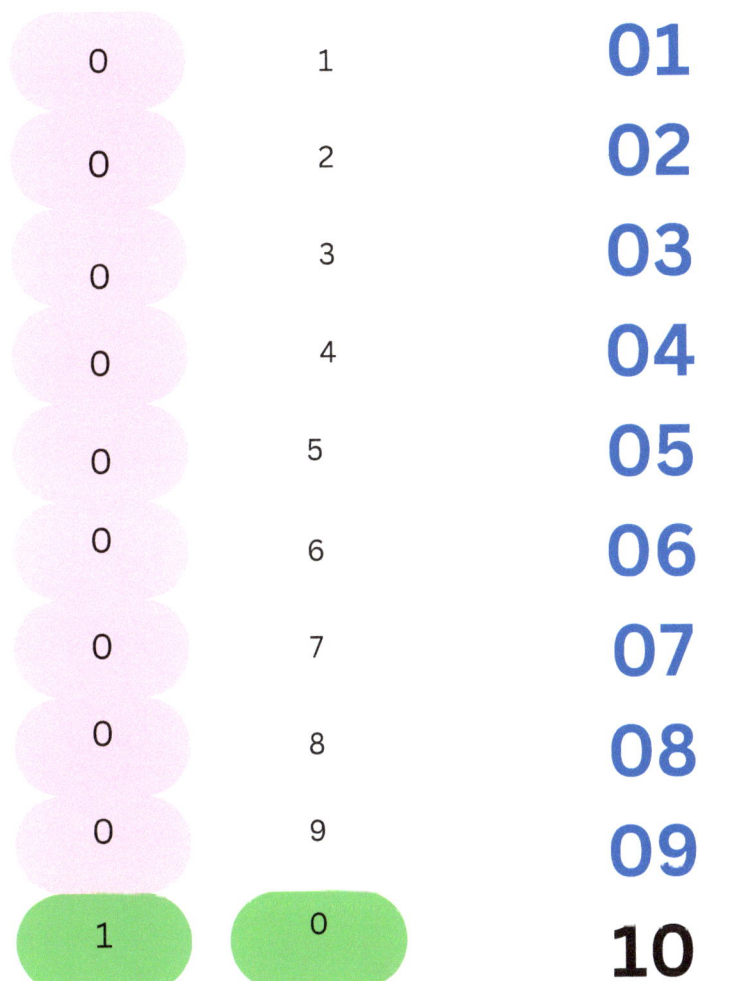

0	1	**01**
0	2	**02**
0	3	**03**
0	4	**04**
0	5	**05**
0	6	**06**
0	7	**07**
0	8	**08**
0	9	**09**
1	0	**10**

Once you reach 9 and add 1 = Starts the process again

Page 69

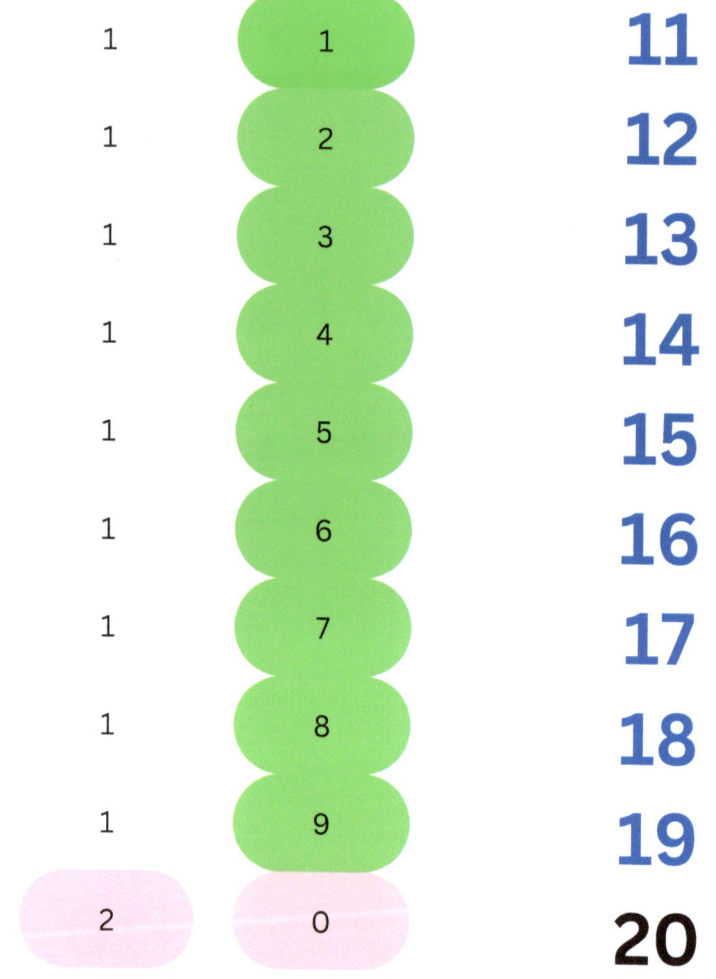

1	1	**11**
1	2	**12**
1	3	**13**
1	4	**14**
1	5	**15**
1	6	**16**
1	7	**17**
1	8	**18**
1	9	**19**
2	0	**20**

Once you reach 9 and add 1 = Starts the process again

Page 70

2	1	21
2	2	22
2	3	23
2	4	24
2	5	25
2	6	26
2	7	27
2	8	28
2	9	29
2	0	30

Once you reach 9 and add 1 = Starts the process again

Page 71

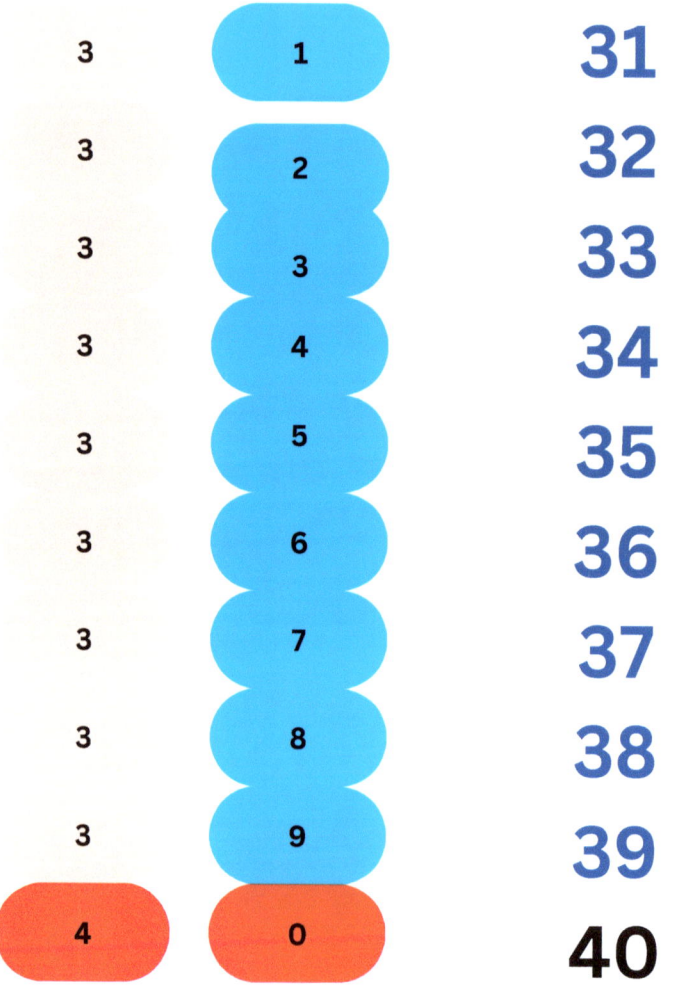

3	1	31
3	2	32
3	3	33
3	4	34
3	5	35
3	6	36
3	7	37
3	8	38
3	9	39
4	0	40

Once you reach 9 and add 1 = 10 Starts the process again

Page 72

4	1	41
4	2	42
4	3	43
4	4	44
4	5	45
4	6	46
4	7	47
4	8	48
4	9	49
5	0	**50**

Once you reach 9 and add 1 = Starts the process again

Page 73

5	1	**51**
5	2	**52**
5	3	**53**
5	4	**54**
5	5	**55**
5	6	**56**
5	7	**57**
5	8	**58**
5	9	**59**
6	0	**60**

Once you reach 9 and add 1 = Starts the process again

Page 74

6	1	**61**
6	2	**62**
6	3	**63**
6	4	**64**
6	5	**65**
6	6	**66**
6	7	**67**
6	8	**68**
6	9	**69**
7	0	**70**

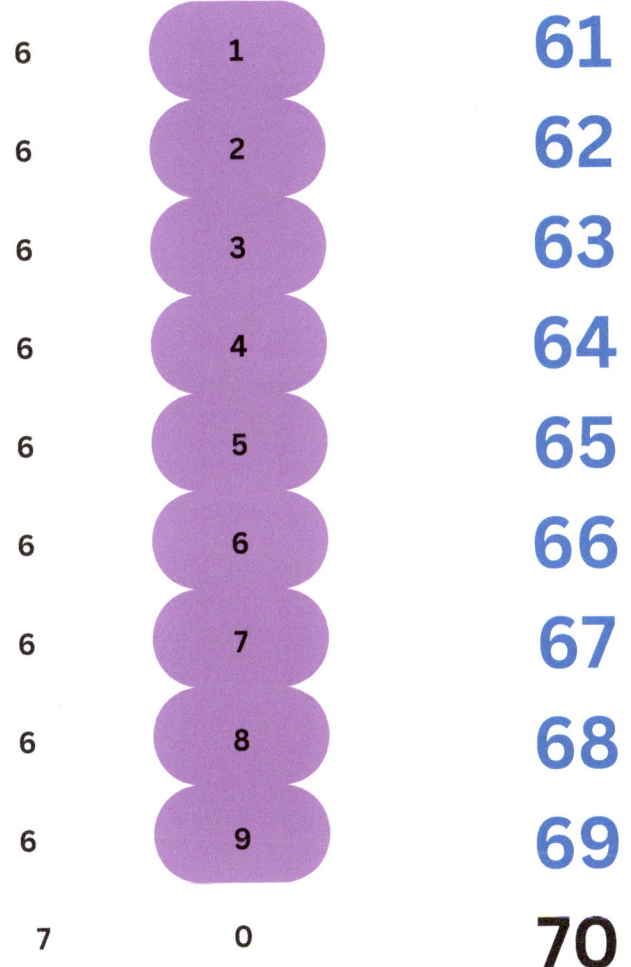

Once you reach 9 and add 1 = Starts the process again

Page 75

7	1	**71**
7	2	**72**
7	3	**73**
7	4	**74**
7	5	**75**
7	6	**76**
7	7	**77**
7	8	**78**
7	9	**79**
8	0	**80**

Once you reach 9 and add 1 = Starts the process again

Page 76

8	1	**81**
8	2	**82**
8	3	**83**
8	4	**84**
8	5	**85**
8	6	**86**
8	7	**87**
8	8	**88**
8	9	**89**
9	0	**90**

Once you reach 9 and add 1 = Starts the process again

Page 77

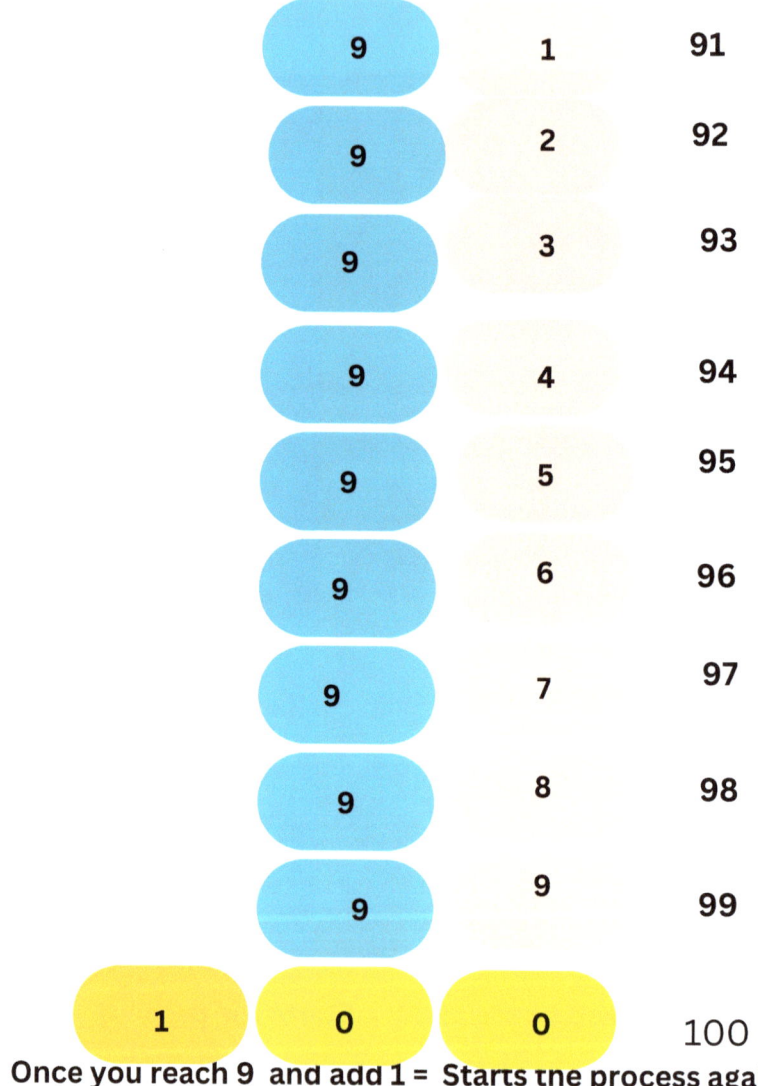

9	1	91
9	2	92
9	3	93
9	4	94
9	5	95
9	6	96
9	7	97
9	8	98
9	9	99
1	0	0	100

Once you reach 9 and add 1 = Starts the process again

Page 78

Counting upto 0- 100 with naming each number

Count 0- 100

zero	0
One	1
Two	2
Three	3
Four	4
Five	5
Six	6
Seven	7
Eight	8
Nine	9
Ten	10

Eleven	11
twelve	12
Thirteen	13
Fourteen	14
Fifteen	15
Sixteen	16
Seventeen	17
Eighteen	18
Nineteen	19
Twenty	20

Twenty-one	21
Twenty- two	22
Twenty -three	23
Twenty Four	24
Twenty-Five	25
Twenty -Six	26
Twenty-Seven	27
Twenty-Eight	28
Twenty-nine	29
Thirty	30

Thirty-one	31
Thirty-two	32
Thrity-three	33
Thirty-four	34
Thirty-five	35
Thirty-six	36
Thirty-seven	37
Thirty-Eight	38
Thirty-nine	39
Forty	40

forty-one	41
Forty- two	42
Forty-three	43
Fourty-four	44
Forty-five	45
Forty- six	46
Forty-seven	47
Forty-eight	48
Forty-nine	49
Fifty	50

Fifty-one	51
Fifty- two	52
Fifty-three	53
Fifty-four	54
Fifty-Five	55
Fifty -Six	56
Fifty-Seven	57
Fifty-eight	58
Fifty-nine	59
Sixty	60

Sixty-one	61
Sixty- two	62
Sixty -three	63
Sixty- Four	64
Sixty-Five	65
Sixty-Six	66
Sixty-Seven	67
Sixty -Eight	68
Sixty-nine	69
Seventy	70

Page 86

Seventy-one	71
Seventy- two	72
Seventy -three	73
Seventy-four	74
Seventy-Five	75
Seventy-Six	76
Seventy-Seven	77
Seventy-Eight	78
Seventy-nine	79
Eighty	80

Eighty-one	81
Eighty-two	82
Eighty-three	83
Eighty-four	84
Eighty-Five	85
Eighty-six	86
Eighty-seven	87
Eighty-Eight	88
Eighty	89
Ninety	90

Page 88

Ninety-one	91
Ninety- two	92
Ninety -three	93
Ninety Four	94
Nienty-Five	95
Ninety -Six	96
Ninety-Seven	97
Ninety-Eight	98
NIney-nine	99
One hundred	100

Count to 100

Page 90

100'S CHART

1	2	3	4	5	6	7	8	9	10
11	12	13	14	15	16	17	18	19	20
21	22	23	24	25	26	27	28	29	30
31	32	33	34	35	36	37	38	39	40
41	42	43	44	45	46	47	48	49	50
51	52	53	54	55	56	57	58	59	60
61	62	63	64	65	66	67	68	69	70
71	72	73	74	75	76	77	78	79	80
81	82	83	84	85	86	87	88	89	90
91	92	93	94	95	96	97	98	99	100

Page 91

Counting by 5

Page 92

Remember the
five
Represents

In Words	Symbol	Count
Five		

Counting by Fives
(Look for the Pattern)

Five	5	5
Ten	5+5	10
Fifteen	5+5+5	15
Twenty	5+5+5+5	20
Twenty-five	5+5+5+5+5	25
Thirty	5+5+5+5+5+5	30
Thrity-five	5+5+5+5+5+5+5	35
Fourty	5+5+5+5+5+5+5+5	40
Forty-Five	5+5+5+5+5+5+5+5+5+	45
Fifty	5+5+5+5+5+5+5+5+5+5	50

Page 94

Page 95

Remember the 10 Represents

In Words	Symbol	Count
Ten	10	

Counting by Tens
Look for the pattern

ten	10	10
Twenty	10+10	20
Thirty	10+10+10	30
Fourty	10+10+10+10	40
Fifty	10+10+10+10+10	50
Sixty	10+10+10+10+10+10	60
Seventy	10+10+10+10+10+10+10	70
Eighty	10+10+10+10+10+10+10+10	80
Ninety	10+10+10+10+10+10+10+10+10	90

Page 98

What is one more than 3?

Step 1 : Count to the third block

1   2   3

Step 2:
Using the number line to identify the next number after 3

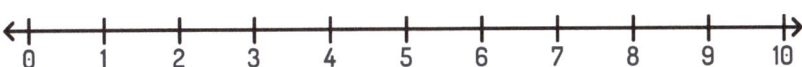

0 1 2 3 4 5 6 7 8 9 10

Step 3 : The next number after 3 is?

1   2   3   4

The number after 3 is 4.
Page 99

One Less

Page 100

What is one less than 3?

Step 1 : Count to the third block

Step 2: Count using the number line to identify the number before 3

Step 3 :The next number before 3 is?

The answer is 2 is before 3.
Page 101

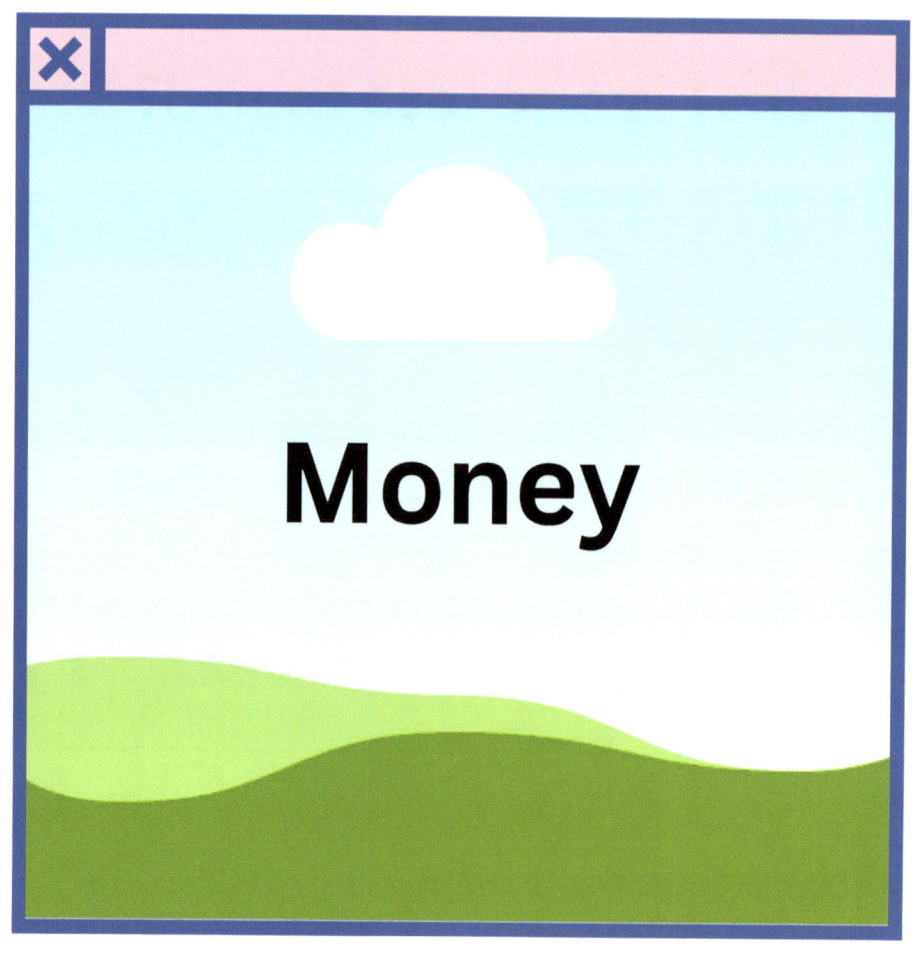

Money

Page 102

	Image	Value	To make a dollar	Create a picture
Pennie		.01	100 pennies	
Nickel		.05	20 nickels	
Dime		.10	10 dimes	
Quarter		.25	4 quarters	

Page 103

Addition

Page 104

What is 3 + 2?

Step 1 : Create a Picture

1 2 **3**
Represents the number three

+
The plus sign means to add

1 **2**
Represents the number 2

Step 2 : Combine all of the items to find out the total

1 **2** 3 4 **5**

Five **is the total number of items**
Step 3 What is 3+ 2 = 5

Page 105

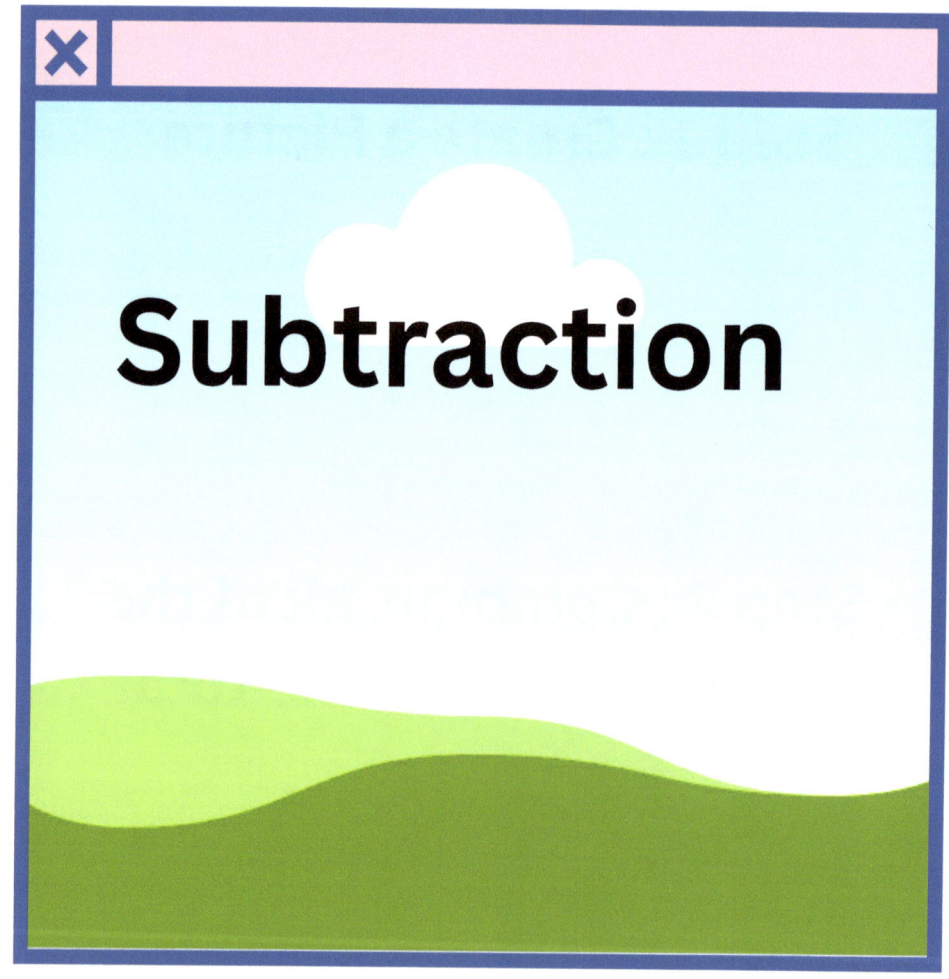

Subtraction

Page 106

What is 3 -2?

3 minus 2

Step1 :   Started with 3 items

Step 2: Minus (Means to take away)

1

Step 3: The final answer is 1

Step: 3-2 = 1[1]

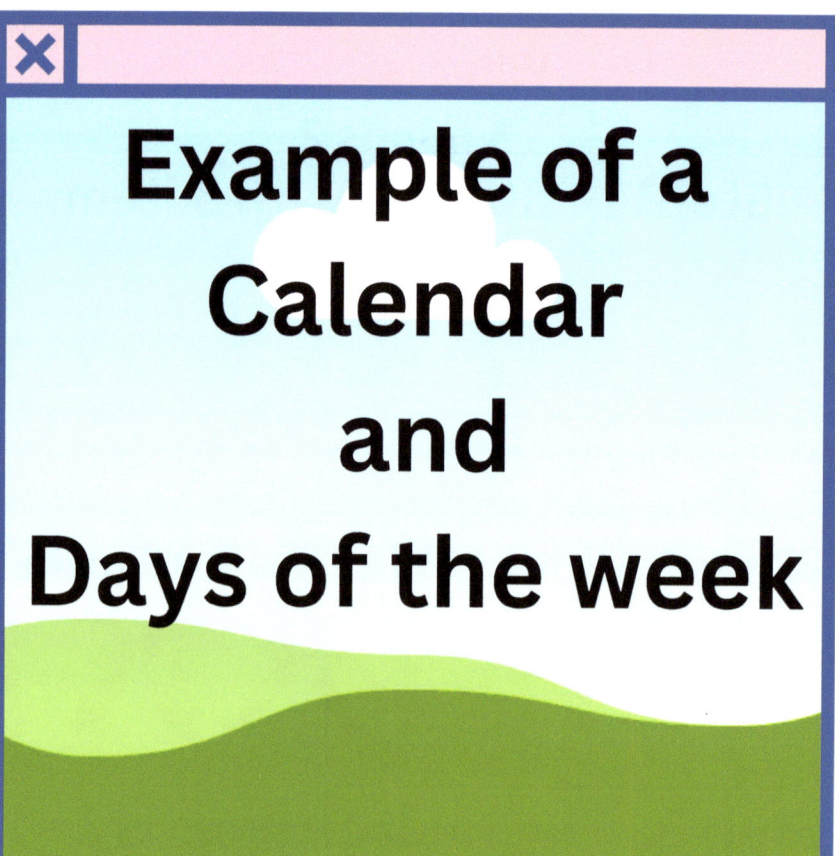

Example of a Calendar and Days of the week

Page 108

days of the Week

Sunday

Monday

Tuesday

Wednesday

Thursday

Friday

Saturday

Page 109

2024

January
M	T	W	T	F	S	S
1	2	3	4	5	6	7
8	9	10	11	12	13	14
15	16	17	18	19	20	21
22	23	24	25	26	27	28
29	30	31				

February
M	T	W	T	F	S	S
			1	2	3	4
5	6	7	8	9	10	11
12	13	14	15	16	17	18
19	20	21	22	23	24	25
26	27	28	29			

March
M	T	W	T	F	S	S
				1	2	3
4	5	6	7	8	9	10
11	12	13	14	15	16	17
18	19	20	21	22	23	24
25	26	27	28	29	30	31

April
M	T	W	T	F	S	S
1	2	3	4	5	6	7
8	9	10	11	12	13	14
15	16	17	18	19	20	21
22	23	24	25	26	27	28
29	30					

May
M	T	W	T	F	S	S
		1	2	3	4	5
6	7	8	9	10	11	12
13	14	15	16	17	18	19
20	21	22	23	24	25	26
27	28	29	30	31		

June
M	T	W	T	F	S	S
					1	2
3	4	5	6	7	8	9
10	11	12	13	14	15	16
17	18	19	20	21	22	23
24	25	26	27	28	29	30

July
M	T	W	T	F	S	S
1	2	3	4	5	6	7
8	9	10	11	12	13	14
15	16	17	18	19	20	21
22	23	24	25	26	27	28
29	30	31				

August
M	T	W	T	F	S	S
			1	2	3	4
5	6	7	8	9	10	11
12	13	14	15	16	17	18
19	20	21	22	23	24	25
26	27	28	29	30	31	

September
M	T	W	T	F	S	S
						1
2	3	4	5	6	7	8
9	10	11	12	13	14	15
16	17	18	19	20	21	22
23	24	25	26	27	28	29
30	31					

October
M	T	W	T	F	S	S
	1	2	3	4	5	6
7	8	9	10	11	12	13
14	15	16	17	18	19	20
21	22	23	24	25	26	27
28	29	30	31			

November
M	T	W	T	F	S	S
				1	2	3
4	5	6	7	8	9	10
11	12	13	14	15	16	17
18	19	20	21	22	23	24
25	26	27	28	29	30	

December
M	T	W	T	F	S	S
						1
2	3	4	5	6	7	8
9	10	11	12	13	14	15
16	17	18	19	20	21	22
23	24	25	26	27	28	29
30						

Thank you for supporting   students. Everyday is a math class and have fun on this journey of discovery.

For more information on creating a student math educational plan.
please contact :
Dr.JPCunlimited@gmail.com
Keep on the lookout for new Math Resources

Page 111